DIG DEEP!
Bugs That Live Underground

Millipedes

Wendell Rhodes

PowerKiDS press

New York

Published in 2017 by The Rosen Publishing Group, Inc.
29 East 21st Street, New York, NY 10010

Copyright © 2017 by The Rosen Publishing Group, Inc.

All rights reserved. No part of this book may be reproduced in any form without permission in writing from the publisher, except by a reviewer.

First Edition

Editor: Sarah Machajewski
Book Design: Mickey Harmon

Photo Credits: Cover (sky) Severe/Shutterstock.com; cover (background) ifong/Shutterstock.com; cover (millipede) davemhuntphotography/Shutterstock.com; pp. 3–4, 6, 8, 10, 12, 14, 16, 18, 20, 22–24 (background) isaravut/Shutterstock.com; p. 5 Kazitafahnizeer/Shutterstock.com; p. 7 Dashu/Shutterstock.com; p. 8 James Steidl/Shutterstock.com; p. 9 (main) zatvornik/Shutterstock.com; pp. 9 (inset), 15 (main) wi6995/Shutterstock.com; p. 11 Noumae/Shutterstock.com; p. 12 https://en.wikipedia.org/wiki/Millipede#/media/File:Rusty_millipedes.jpg; p. 13 © Custom Life Science Images /Alamy Stock Photo; p. 15 (nymphs) https://commons.wikimedia.org/wiki/File:Chordeumatida_sp._Nanogona_polydesmoides.jpg; p. 15 (eggs) https://upload.wikimedia.org/wikipedia/commons/2/2f/Brachycybe_with_eggs.JPG; p. 16 tcareob72/Shutterstock.com; p. 17 jamesteohart/Shutterstock.com; p. 19 skynetphoto/Shutterstock.com; p. 21 Matteo photos/Shutterstock.com; p. 22 TUM2282/Shutterstock.com.

Cataloging-in-Publication Data

Names: Rhodes, Wendell.
Title: Millipedes / Wendell Rhodes.
Description: New York : PowerKids Press, 2017. | Series: Dig deep! bugs that live underground | Includes index.
Identifiers: ISBN 9781499420586 (pbk.) | ISBN 9781499420609 (library bound) | ISBN 9781499420593 (6 pack)
Subjects: LCSH: Millipedes–Juvenile literature.
Classification: LCC QL449.6 R494 2017| DDC 595.6'6–dc23

Manufactured in the United States of America

CPSIA Compliance Information: Batch #BS16PK: For Further Information contact Rosen Publishing, New York, New York at 1-800-237-9932

Contents

Down on the Ground 4
Many, Many Millipedes. 6
Thousand-Footed 8
Millipede Meals 10
Eggs in the Soil 12
Baby Millipedes 14
Proper Protection 16
Not a Pest 18
Millipedes vs. Centipedes 20
Millipedes Are Cool! 22
Glossary 23
Index. 24
Websites 24

Down on the Ground

What's long and thick with hundreds of legs? It's the millipede, one of the creepiest, crawliest bugs around! Millipedes spend their life in or on the ground, hiding in leaves and eating dead plant matter. However, it's their life on the ground that sometimes leads them into our homes—especially the basement!

Millipedes can't harm us, but you might think they're pretty gross. They're really big for bugs and they crawl rather slowly. Digging deeper into the life of millipedes teaches us some cool facts about them. Maybe they're not as icky as you think!

This millipede is on the move!

5

Many, Many Millipedes

Millipedes are very common creatures, and there are plenty of them in the world. Scientists know of at least 10,000 different species, or kinds, of millipedes. There are probably many more. The common North American millipede is one of many species that live in the United States.

Millipedes live all over the world, and you can find them in most **habitats**. They live on land, in soil that's moist, or wet. **Tropical** habitats are home to many millipede species. However, you can commonly find them in fields and in the woods. Millipedes can make their home anywhere as long as it's dark and **damp**.

The woods are home to many creepy, crawly creatures—including millipedes!

7

Thousand-Footed

If millipedes are known for one thing, it's probably their legs. There are a lot of them. Their name means "thousand-footed." While millipedes don't actually have a thousand legs, some species have as many as 200 pairs.

Millipedes aren't **insects**, although they belong to the same animal group—arthropods. Millipedes' bodies are long and segmented, which means they're split into many sections. The first three or four segments each have one pair of legs, while most of the rest have two pairs. Millipedes also have antennae, or feelers, and mouthparts they use to chew. A hard outer covering called an exoskeleton protects a millipede's body.

Dig Deeper!

The giant African millipede can be up to 12 inches (30.5 cm) long! Its body can have between 30 and 40 segments.

mouthparts

feelers

legs

exoskeleton

segment

Millipedes have tiny holes called spiracles all over their body. They use them to breathe.

Millipede Meals

Most millipedes are herbivores. That means they eat plants. They're also **decomposers**. That means they feed on **decaying** matter. That includes dead leaves, wood, and other plant parts. Damp, dark places are perfect for finding this kind of food. If their habitats get too dry, millipedes will feed on healthy, green leaves. These leaves are a good source of water.

Some millipedes feed on partly decayed animals, including worms, snails, and bugs. This keeps habitats healthy. Millipedes are mostly nocturnal, which means they're active at night. That's when they eat!

As millipedes eat, they release nutrients back into the soil. This keeps their habitats healthy.

Dig Deeper!

Millipedes have bacteria and other **microbes** in their body that help them break down their food.

Eggs in the Soil

Millipedes begin life as an egg. Male and female millipedes come together to **mate**. Then, the female millipede lays its eggs. Different species lay their eggs in different places. Some millipedes lay their eggs underground, while others lay them under logs or leaf matter.

The North American millipede does something interesting with its eggs. After mating, it eats soil, which passes through its body. The millipede shapes the matter into a nest and lays her eggs in it! Some millipedes guard their nests carefully until the eggs hatch. Not all species do this, though.

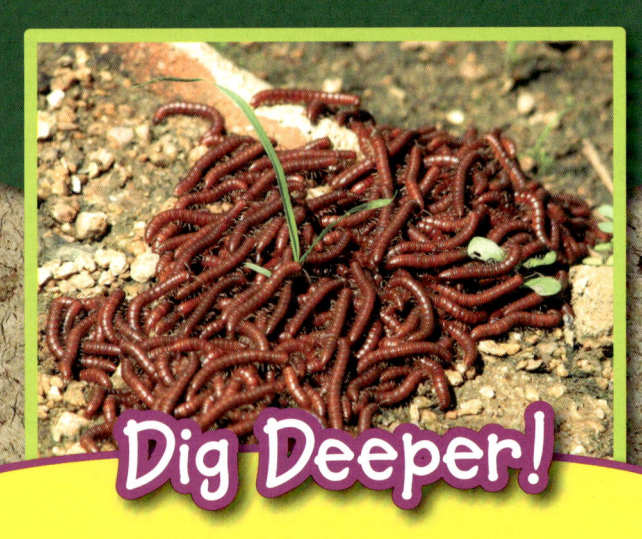

Dig Deeper!

Groups of millipede eggs are called clutches.

After mating, females lay between 20 and 300 eggs at a time.

Baby Millipedes

Baby millipedes are called nymphs. They look a lot like adult millipedes. However, their bodies are white and kind of clear. They're born with segmented bodies. They have three to four pairs of legs, depending on the species.

Millipedes get bigger as they grow into adults. They molt several times as they grow. Molting is when the millipedes shed their exoskeleton, which creates room for the millipedes to grow bigger. As they grow, they gain more segments and more pairs of legs. Millipedes are adults when they have all their segments and legs.

A millipede's life cycle has three stages—egg, nymph, and adult.

adult

nymph

egg

Proper Protection

Even though they have hundreds of legs, millipedes can't run very fast. That means they can't quickly escape from predators. They can't bite or sting either. Luckily, millipedes have other ways of **defending** themselves.

Millipedes curl up when they sense danger. Their legs are safely tucked away, while the hard exoskeleton keeps predators from hurting them. Some millipede species have a pretty gross defense. Their bodies can give off a bad-smelling liquid or gas. This tells predators to stay away! One kind of millipede glows in the dark. This warns predators that its body contains poison.

Spiders, ants, beetles, frogs, birds, and lizards are known millipede predators.

Not a Pest

Millipedes live everywhere, even near people. It's common to find them under piles of leaves or grass clippings. If you turn over a flowerpot or dig in your garden, you might see a millipede crawling through the soil.

Millipedes are known to leave their homes in the fall or after it's rained a lot. This is commonly when they show up in our homes, especially basements. If you see one, don't worry. Millipedes don't harm people. And they can't live very long in dry places. If you see a millipede inside, simply take it back outside. Picking it up won't hurt you.

Dig Deeper!

A pest is a creature that can cause harm to people, homes, or crops.

Millipedes may look scary, but they're actually harmless. Don't be afraid of them!

Millipedes vs. Centipedes

Millipedes are often grouped with another arthropod you're probably familiar with—centipedes. These creepy crawlers look and act a lot like millipedes, but there are some key differences.

Centipedes and millipedes both have segmented bodies, though centipedes are generally bigger. Centipedes have many legs, but not as many as millipedes. Their name means "hundred-footed." They can have as few as 14 pairs of legs or as many as 177 pairs.

Like millipedes, centipedes live in dark, moist habitats. They feed on decaying plant matter, too. However, centipedes don't curl up when they're **threatened**. They eat meat, and they have fangs to bite their **prey**.

Centipedes and millipedes are a lot alike. Can you tell the differences between them in this picture?

centipede

millipede

Millipedes Are Cool!

Millipedes are considered beneficial creatures. That means they help their habitat. As decomposers, they keep their habitat healthy. They get rid of the dead stuff that would otherwise be lying around!

Millipedes are most active at night, so you may not have many chances to see them. However, they'll crawl away if you find the pile of leaves or soil they're hiding in. And you probably won't see them at all during the winter. Adults spend this season underground. When they come above ground in spring, they get back to eating and crawling. Millipedes are cool!

Glossary

damp: Slightly wet.

decay: To rot.

decomposer: A living creature that eats dead plant or animal matter.

defend: To keep safe from harm or danger.

habitat: The natural home for a plant, animal, or other living creature.

insect: An animal that has six legs and one or two pairs of wings.

prey: An animal hunted by another animal for food.

mate: To come together to make babies.

microbe: A creature so small it can't be seen without a microscope.

threatened: To be put in danger.

tropical: Having to do with areas that have hot, humid habitats.

Index

A
ants, 17
arthropods, 8, 20

B
beetles, 17
birds, 17

C
centipedes, 20, 21

E
eggs, 12, 13, 14, 15
exoskeleton, 8, 9, 14, 16

F
feelers, 8, 9
frogs, 17

G
giant African millipede, 8

L
legs, 4, 8, 9, 14, 16, 20
lizards, 17

M
microbes, 11
mouthparts, 8, 9

N
nocturnal, 10
North American millipede, 6, 12
nymph, 14, 15

P
predators, 16, 17
prey, 20

S
snails, 10
species, 6, 8, 12, 14, 16
spiders, 17
spiracles, 9

U
United States, 6

W
worms, 10

Websites

Due to the changing nature of Internet links, PowerKids Press has developed an online list of websites related to the subject of this book. This site is updated regularly. Please use this link to access the list: www.powerkidslinks.com/digd/mill